The Seniors' Survival Guide

New tricks for old dogs

Geoff Tibballs

The Seniors' Survival Guide

New tricks for old dogs

MICHAEL O'MARA BOOKS LIMITED

First published in Great Britain in 2008 by
Michael O'Mara Books Limited
9 Lion Yard
Tremadoc Road
London SW4 7NQ

A CIP catalogue record for this book is available
from the British Library.

ISBN: 978-1-84317-236-9

1 3 5 7 9 10 8 6 4 2

www.mombooks.com

Cover design by Oil Often

Text design and typesetting by Ana Bjezancevic

All images from www.shutterstock.com, except pages 56, 108, 128 and
130 from www.iStockphoto.com, and pages 7, 8, 10, 16, 24, 61, 63, 64,
65, 68, 71 and 144 © 2008 www.clipart.com.

Printed and bound in Great Britain by Clays Ltd, St Ives plc

CONTENTS

❯ Introduction   7
❯ Text Messages   11
❯ The Internet   15
❯ PINs and Other Security Codes   24
❯ Satellite Navigation Systems   27
❯ Teenspeak   33
❯ Call Centres   40
❯ Email   47
❯ Starbucks   52
❯ Mobile Phones   55
❯ Flat-Pack Furniture   61
❯ Digital Cameras   66
❯ Cash Dispensers and
    Other Automatic Machines   68
❯ Global Warming   74
❯ Facebook, Friends Reunited
    and Other Such Things   80
❯ Hi-Tech TV Sets   86
❯ Crime   91
❯ Booking Holidays Online   97
❯ Washing Machines   102
❯ iPods   105
❯ Telephone Answering
    Machine Messages   110

❯ Modern Terminology,
    Including Political Correctness   114
❯ Digital Clocks and
    Video Recorders   126
❯ Banks with Real People
    in Them   132
❯ Electronic Garden Gadgets   139
❯ Reliable Tradespeople   144
❯ Buying and Selling on eBay   149
❯ Video Games   156
❯ Dressing Your Age   160
❯ Budget Airlines   166

INTRODUCTION

I always thought the idea of reaching 'a certain age' was that we would be able to lean back in our reclining chairs and bask in the wisdom and knowledge accrued over the years, occasionally imparting the odd nugget to a grateful younger generation. So much for theory, because instead of us teaching them, it's them teaching us! Watch any eight-year-old sending a text message and he or she can do it in less than a minute. Ask a mature person to send the same text message and it takes us the entire *EastEnders* omnibus, and occasionally runs over into *Songs of Praise*.

Modern technology can be so bewildering. For a start, there's so much of it – LCD, LED, PDA, DAB, DVD... FFS! In our day, advanced technology began and ended with whether or not we could get a signal from Radio Luxembourg. Now no sooner have we mastered one skill than that's out of date and there's something new to get our heads around. It's little wonder we look old. Digital this, digital that, PINs and pods, electronic timers that you have to keep programming, SatNav, mobile phones, not to mention anything to do with the ruddy Internet. And why do banks, airlines, utility companies and so on automatically assume that everybody has an Internet connection? Did I miss the day it was being given out free?

The major reason why modern technology is so baffling to us is not that it is necessarily complicated in itself, but because it is wrapped in such incomprehensible jargon. *The Da Vinci Code* is easier to unravel than most instruction manuals. It is hardly surprising, therefore, that a recent survey among adults revealed that one in five gadgets remains unopened in the box. At least there it can't hurt us.

But it's not only gadgets that make life today so frustrating. Try finding a bank with real people in it, a company that doesn't have one of those wretched call centres on the sub-continent, or understanding a single word that a teenager says after you have politely asked him not to stand on the bonnet of your car. Youngsters have a language all of their own, a trait they unfortunately share with most people in middle management who talk about singing from the same hymn sheet, pushing the envelope, touching base, and running it up the flagpole. Surely not all at the same time?

There are so many things to remember these days – weighing letters before you post them, security codes, passwords, which utility or phone company you're with, and so on. It's not helped by the fact that none of these companies seems to have a sensible name: npower, T-Mobile, onetel, E.ON, EDF, O2. Oh, for the days when firms had proper names like Brown & Polson, Crosse & Blackwell or Mr Whippy.

There are two ways of dealing with modern life. Either you can try to ignore it and hope that it will go away – rather like some people did when Stephenson unveiled his *Rocket* and when decimalization was introduced – or you can try to pick up the odd skill without showing yourself up in front of the younger generation. You don't have to embrace new technology as such; a peck on the cheek will suffice.

To this end, you may find the occasional piece of useful advice in this book – although if you do, it will be more by luck than design. For if you don't know your WiFi from your HiFi, your BlackBerry from your Burberry, Bluetooth from *Blue Peter*, or a JPEG from a clothes peg... join the club.

TEXT MESSAGES

Texting: A step-by-step guide

❶ Start writing a message using the number pad, which, for the purposes of this task, is now a letter pad.

❷ Mistype a word and accidentally delete the whole message while trying to backtrack.

❸ Start again.

❹ Accidentally send a half-written message to the wrong person while trying to find the question mark.

❺ Start again.

❻ Eventually cobble together some sort of missive, only to discover that it's five characters too long to fit into one message.

❼ Mutter 'I'm buggered if I'm going to waste another 12p on a second text' and go through the message deleting all punctuation and inserting words such as '2mz'[1] in order to cut it down.

❽ Send, and then pour yourself a stiff drink.

---

1 Short for '2moz', which is short for '2moro', which is short for 'tomorrow'.

Txt-spk

Txt-spk is a thoroughly impenetrable language reserved for communication via text message. While it may look like a series of algebraic formulae, txt-spk is in fact based on everyday English, simply massacred in order to save space (and thus money) when texting.

Mastering txt-spk means throwing all care for punctuation and grammatical correctness out of the window, so, if you are rather a language pedant, steel yourself.

ur	your, you're, you are
da	the
n	and
btw	by the way
imho	in my humble/ honest opinion
gr8	great, as in 'gr8 2 c u' (8 is a handy number in txt-spk: sk8, w8, infuri8 etc.)
wkd	wicked, as in gr8
lol	laughing out loud, as in 'ur so wkd, lol'
lmao	laughing my arse off
roflmao	rolling on the floor laughing my arse off (I kid you not)

It goes on n on n on but, imho, listing da whole lot would take 4eva.

The perils of predictive texting

Mobile phones are idiots. There you are sending what you think is a perfectly straightforward message, and the blasted machine second-guesses you and spouts forth all manner of nonsense, all in the name of labour-saving.

On the plus side, predictive texting is quite handy for the wilfully lazy, as you can pretty much write whole narratives by repeatedly pressing the same sequence of buttons. Assuming, that is, that you want to write about 'any coy cow' or an 'aloof, blond clone, alone'.

Some phones go so far as to choose 'nun' over 'mum'. While today's nuns are no doubt more 'down with the kids' than nuns of yore, the chances of you sending a text message about one are surely quite low...

The golden rules of text messaging

It's a minefield out there, but if you stick to these tenets you should survive relatively unscathed.

1 Don't expect too much too soon. When you watch kids, their fingers and thumbs are a blur, but, initially at least, you'll probably be about as adroit at texting as Captain Hook. Indeed, your speed will probably be such that it would be quicker to send the message by an arthritic pigeon with one wing. Accept this.

2 Learn some txt-spk, even if it is through gritted teeth. When we went to school, vowels were still considered part of the English language, but times change. Youngsters save time by omitting vowels and other letters they don't like – a practice that some then carry on into GCSE exam papers and wonder why they get an F grade.

3 Have a teenage interpreter on permanent standby.

THE INTERNET

Everyone's always going on about the Internet. They say it's the best thing since sliced bread, but nobody ever stayed up all night gazing at a packet of Wonderloaf. So there must be something to it. You can do just about anything on the Internet – it's even more versatile than a Swiss army knife – but is it safe? You hear such awful stories about these chat rooms. A lot of people say that wherever you go on the Internet there's porn, but this is not strictly true. You have to know where to look for it.

What does it do, then?

One of the biggest problems is where to begin, so here are a few interesting sites to help you on your way:

> PostSecret.com is a collection of anonymous and often rather salacious secrets sent in on picture postcards.

> MauriceBennett.co.nz illustrates the New Zealander's unique ability to create works of art from toast. His portfolio of lifelike toast portraits includes Elvis Presley, the Mona Lisa and Dame Edna Everage.

> MyCatHatesYou.com promises 'the largest collection of sour-faced, indignant felines on the Internet'.

> BoxWars.co.uk is *the* site for people who meet in order to fight majestically choreographed battles wearing suits of armour made from cardboard.

> TomatoesAreEvil.com is a site dedicated to the belief that the humble tomato is in fact evil.

> BedBugger.com offers tips on how to combat bed bug infestation.

> AirToons.com presents a vast collection of parodied airline safety cards, lovingly gathered from around the world.

Googling

Finding what you're searching for on the Internet is like looking for a needle in an entire field of haystacks. So some clever person invented search engines, of which the most popular is Google, to help us floundering fools track down the information we are after. Simply type in the keywords and the most appropriate sites will be displayed... If only it were that simple.

Say that, as a proud native of Yorkshire, you want to find out something about the history of Hull and type in the words 'Hull history'. Suddenly you'll be confronted with page after page of sites about Hull, Alabama – and ten other American Hulls; the hull of the *Titanic*; the Hull-White model of interest rates; the careers of Rod Hull and Emu; and hiding away in the middle somewhere will be Hull, UK.

It is certainly advisable to think carefully before keying in your search words. An innocent request for bush clippers can lead you into unwelcome areas, and for information on the Bond character Pussy Galore it's probably best to go to your local library.

THE GOLDEN RULE
When in doubt, think of your computer as
a thirteen-year-old schoolboy.

Pop-ups

These are the Internet equivalent of those people who knock on your door unannounced and try to sell you dusters, tea towels and God. If only you could get rid of pop-ups by setting the dogs on them.

The trouble with pop-ups is not only are they a nuisance but they can also infiltrate your computer, slowing it down and causing you to get unsettling messages on your screen, ranging from 'WARNING: YOUR COMPUTER MAY BE INFECTED' to the even more alarming 'A FATAL ERROR HAS OCCURRED'. Not serious: fatal.

In such circumstances, the best advice is to close down your computer, unplug it and hope that whatever was upsetting it has gone by the morning. If that doesn't work, try gently talking to it and telling it how much you value its companionship. If that too fails, take an axe to it.

Should anything go wrong, technical advice is always at hand.

You can phone a helpline if you have nothing better to do for four hours and don't mind being treated like a single-cell organism. The whiz-kid at the other end of the line rattles off a string of technical terms (such as 'IP address', 'scart plug' and 'encryption protocols'), which, for all the sense they make, could be part of the language of the planet Zarg.

Then, when you politely inform him that you haven't a clue what he's talking about, he emits an audible sigh of exasperation, mutters something derogatory to his colleague, and asks 'Have you tried switching it off and on?'

Talking the talk

Naturally, Internet vocabulary is virtually impenetrable but here are a few words you might come across:

Bandwidth: This is how much stuff you can send through a connection. Apparently a fast modem can move around 57,000 bits per second, which would sound impressive if you knew what a bit was.

Blog: This is a sort of Internet diary. Someone who keeps a blog is a blogger, although curiously someone who keeps a mug isn't necessarily a mugger.

Browser: Software that helps you look around the Internet. What do you mean, what's software?

Cookie: A piece of information that is sent to your browser for no obvious reason whatsoever and, worse still, doesn't contain any bits of chocolate.

Download: See **upload**.

Hit: A recorded visit to a website. Thus Michael Bolton's website has probably had more hits than he has this year. Actually, one visit would do that.

JavaScript: Some fancy programming language that invariably fouls up your computer.

Log in: Connect to a computer system. In order to log in, you need a password. Anything will do, although it may be best to avoid 'Osama2'.

Modem: A device that connects your computer to your phone line, and is thus fairly important.

Software: The various programs that go on your computer. Confusingly, software programs are on disks, which are quite hard.

Spyware: Secretly installed software that cunningly spies on you to see which sites you visit so that it can bombard you with advertisements that you don't want.

Upload: See **download**.

Virus: Nasty disease that gets into your computer and can eventually cause it to crash. When your computer crashes, it suddenly shuts down and you lose everything on it. This is not a good thing.

Web: Another name for the Internet. It's called the Web because, like a hapless fly, once you're on it, you can't get off.

Wilfing: 'WILF' stands for 'What was I Looking For?' and applies to people who surf the Internet without any real purpose – which applies to most of us, really.

Whether or not the Internet really is the best thing since sliced bread is debatable. One thing is certain, however: you will become so frustrated by it that at some point you will be sorely tempted to put your start-up disk into the toaster just to teach it a lesson.

PINS AND OTHER SECURITY CODES

Once upon a time, we paid for items by cash. It was a perfectly satisfactory arrangement: we handed over the money and received the goods in return. Simple.

You can't help thinking that 'Chip and PIN' was invented by the same people who came up with the solar-powered torch and the chocolate teapot. Surely there are considerably fewer competent forgers in this world than there are people capable of looking over your shoulder while you key in your PIN down the local?

Sod the rules – we're incompetent

We're told to memorize our security codes and passwords, but it's simply impossible. The only thing for it is to write them down, but where? Burglars raised on old episodes of *Dixon of Dock Green* will instinctively look in the biscuit tin for all household valuables, so avoid hiding your details there, unless the slip of paper is cleverly concealed within the layers of a custard cream.

Some innovative places to write down codes:

> ❯ On a scrap of paper hidden in a locket or pocket watch
> ❯ Etched onto the inside rim of your glasses
> ❯ Woven into the pattern of your dress/tie
> ❯ In graffiti beside your favourite hole in the wall
> ❯ Tattooed upon your person
> ❯ Tattooed upon the forehead of your favourite check-out attendant

Places not to write your security codes:

> Anywhere near the words 'BURGLAR ALARM CODE' or 'IMPORTANT SECURITY NUMBERS'
> In dust on the window
> In the classified ads section of your local newspaper
> In spray paint on the wall of your lounge; no reputable burglar will be fooled into believing it is modern art

Got your number

To make it easier to remember your codes, arrange for your various bank accounts to have the same four-digit PINs. The most popular – and obvious – choice for security code numbers is a birthday. If this is the only way you'll remember it, be more inventive by choosing your cat's birthday rather than your own. Any burglar who can work that out almost deserves to make off with your DVD player.

P.S. To any burglar or fraudster reading this book, none of my security codes is my cat's birthday.

SATELLITE NAVIGATION SYSTEMS

It is estimated that within the next ten years, ninety per cent of cars will have satellite navigation systems. The manufacturers would have us believe that a satellite navigation system is an indispensable tool for today's motorist, just as a set of furry dice was in the 1970s.

But surely the boom in SatNav is bad news for those who practise the noble and underrated art of map reading.

An ode to map reading

The joy of being guided down a three-mile-long single-file country lane with no passing places... and coming face to face with a large tractor.

The thrill of being able to know in advance whether the church in the next village has a tower or a spire and whether the forest half a mile away is coniferous or not.

The shared laughter as you calmly point out to your wife that the meandering blue motorway in Berkshire that she has just tried to take you down was not the M4 but the River Thames.

Now it seems that road maps have gone the way of vinyl LPs, rag-and-bone men, Spangles and the career of Keith Chegwin. But a recent survey revealed that four out of every ten people with SatNav have either got hopelessly lost or seriously delayed as a direct result of using it.

> In Exton, Hampshire, dozens of lorry drivers found themselves stuck in a lane that was only six feet wide after following SatNav instructions. As a result the village erected signs telling drivers of heavy goods vehicles to ignore their computers' directions and turn back.

> An ambulance crew transferring a patient on what should have been a ten-mile journey between two hospitals in Essex ended up driving 200 miles in the wrong direction after relying on SatNav. It was only when they reached Manchester that they realised they were 'a little lost'. The thirty-minute journey eventually took over nine hours.

> A taxi driver drove two girls eighty-five miles in the wrong direction after keying the wrong place name into his SatNav. The girls had asked to go from Bournemouth to Lymington, Hampshire, but ended up in Limington, Somerset. After being dropped off, they knew they were in the wrong place when they asked bemused locals for directions to their New Forest campsite.

> In November 2006, a woman dodged oncoming traffic for fourteen miles after misreading her SatNav system and driving the wrong way up a dual carriageway.

> One woman's satellite navigation system ordered her to turn left at a railway level crossing, so she did just that... and ended up perched on the Brighton-to-Hastings line. Ironically, she had once been a member of a committee aimed at reducing road casualties.

> In December 2007, fifty members of a Gloucester-shire social club en route to a Christmas market in Lille, France, ended up in Lille, Belgium – seven hours and one hundred miles away – when their coach driver relied on SatNav for directions. They returned empty-handed and rather lacking in Christmas cheer.

As someone very wise once said, 'To err is human, but to really foul things up you need a computer.'

The rules of SatNav survival

❶ Exercise a degree of common sense. It may tell you to take the next left, but if the next left is clearly someone's driveway, don't take the instructions too literally.

❷ Find one that talks in miles. It's no good being told to turn right in nought-point-three kilometres. What's nought-point-three of a kilometre when it's at home? By the time you've converted it into yards or fractions of a mile, you've missed your turning.

❸ If TomTom drives you madmad, remember: there is always a solution in the glove compartment. It is called a map and has many advantages, not least the fact that SatNav would never be able to tell you when you're approaching a good tumulus.

TEENSPEAK

Teenagers, right? What are they like?! You ask them a perfectly civil question, such as 'Would you mind playing somewhere else with your crossbow?', and you're either met with a volley of abuse, an indecipherable grunt, a knowing silence or 'Talk to the hand 'cos the face ain't listening.' And that's just the ones who work in customer services departments.

The trouble with teenagers is that they think they invented the term, whereas in fact *we* were the first teenagers, and we had our own language – with trendy words such as 'fab', 'cool' and 'groovy'. And we had teenage angst ages before today's lot had their first strop. What have they got to sulk about, anyway?

Things the youth of today should be grateful they don't have to deal with:

Bread and dripping	Outside toilets
Cilla Black	The nit nurse
Coal tar soap	Meat paste
Semolina	Maggots in apples
Latin	NHS specs
Corporal punishment	Margaret Thatcher

Kids today! They don't know they're alive!

Q: How can we bridge the generation gap if the simple act of taking an interest in teenagers' music and fashion suddenly renders those things entirely uncool?

A: Don't bother trying to befriend a teenager. As far as they're concerned, if you're older than twenty you're practically dead. The key is to observe their lingo and pepper your chat with a few 'phat' words, thus proving beyond doubt that you are 'down with the kids'. Or whatever it is they say these days.

Watch and learn: A role-playing exercise

Find a friend to help you with this simple exercise. One of you will play the clueless old duffer (i.e. yourself) and the other will play a modern youth. In this scenario, the acne-ridden youth is behind the till at a sports shop, doing his best to avoid serving anyone.

Duffer: 'Are these £59.99 or £49.99? The ticket isn't clear.'

Youth: 'Dunno.'

Duffer: 'Do you have them in a size 11?'

Youth: 'Not sure.'

Duffer: 'Might there be some in the stock room?'

Youth: [shrugs]

Duffer: 'Well, perhaps I'll have to take my custom elsewhere.'

Youth: 'Pff... Whatever.'

Duffer: 'Do you mind if I hit you over the head with a block of wood?'

Youth: 'I'll have to ask the manager.'

The teenspeak dictionary

So that you can understand the occasional word when teenagers get together on the back seat of the bus and embark on one of those quick-fire, talk-over-each-other diatribes that sound like something off *The Jeremy Kyle Show*, here is a helpful guide to teenspeak:

Bait: Nothing whatsoever to do with fishing, 'bait' in teenspeak means 'blindingly obvious'. ('Didn't you know? That's so bait.')

Bare: A lot of, very. ('He's bare fit' means 'He's very good-looking.')

Beast: Someone who excels at something. ('Norman Hunter was a beast at football.') Also: an unattractive woman.

Bling: Jewellery. ('He's wearing so much bling, he rattles when he walks down the street.')

CBB: Abbreviation for 'can't be bothered'. ('No, I'm not coming. I've got a severe case of CBB.')

Chirp: To chat up. ('We chirped some fit birds last night.')

Clappin': Out-of-date, usually referring to clothes. ('I gotta go shoppin'. My trainers are clappin'.')

Crump: Really bad. ('That zit on your neck, man, it's crump.')

Dred: Dreadful, terrible. ('Where'd ya get ya hair done? Them's dred locks.')

Dry: Stupid, boring, unfunny. ('That's the worst joke I've ever heard. It was so dry.')

Flat roofin': To be overworked and stressed. ('Sorry I couldn't make last night – I was flat roofin' for my GCSEs.')

Fudge: An extremely stupid person, F, U, D, G and E being their predicted GCSE grades. ('She's so fudge, her dog teaches her tricks.')

Greebo: Someone who listens to rock or heavy metal music and doesn't follow fashion trends. ('You don't wanna be goin' out with a greebo!')

Hangin': Describes someone who has poor dress sense. ('See that bitch over there. She's hangin'.')

Kickass: Cool, great. ('Check out this kickass website.')

Killer: Brilliant, wonderful. ('That was a killer party on Saturday.')

Lame: Weak, feeble, pathetic. ('Jordan sucks. She's so lame.')

Lush: Good-looking. ('That guy on the bike is so lush.')

Minging: Ugly. ('I've never seen anybody so minging!' Hence 'She's got a face that could stop clocks – she's a right minger.')

Mint: Cool. ('Love the shades. They're mint.')

Pants: Not very good. ('I love Wayne's shirt but his jeans are pants.')

Redick: Ridiculous, unfair. ('Why should I do my homework? That's so redick!')

Roll with: Hang out with. ('I'm rollin' with Jenny tonight. OK?')

Safe: Good. ('Yeah, five o'clock is fine, safe.')

Shabby: Smart. ('That suit's well shabby.')

Swag: Scary. ('Did you see that fight outside KFC? It was well swag.')

Sweet: Excellent, awesome. ('That guy who climbed the Eiffel Tower. That was so sweet.')

TTM: Superlative, an abbreviation for 'To the Max'. ('Wow, this game is awesome, TTM!')

Wicked: Cool. ('Don't ya fink Girls Aloud's latest song is wicked?')

So when teenagers moan that nobody understands them, they're right. Nobody *does* understand them. Maybe they should try using words like 'fab' and 'groovy'. Then we could all be cool cats together.

CALL CENTRES

It used to be that, if you had a couple of hours to kill, you would go for a walk, read a book or potter about in the garden. These days you can spend the same amount of time in a less pleasurable pursuit: struggling to get through to a call centre, usually located thousands of miles away.

Indian operators are unfailingly polite but you can't expect a charming young man in Pondicherry to be any more familiar with the stations on the Liverpool Street to Colchester line than you would be with the names of dry cleaners in Hyderabad. The inevitable result is that your query remains spectacularly unanswered, rendering the entire process utterly pointless.

Always remember: If you phone a call centre and a voice at the other end answers within seconds, do not make the mistake of thinking you are getting somewhere. This is merely a machine about to bombard you with a long list of options, and is only the first annoyance on a long journey of woe and despair.

Numbers game

The joy of call centres is that you have options. Oh, so many options. More options than are rationally possible for your very simple query:

Press '1' if you are calling about your water meter reading.

Press '2' if you would like to have a water meter installed.

Press '3' if you are calling about your water bill.

Press '4' if you are calling about the quality of your water.

Press '5' if you are calling to report a leak.

Press '6' if you're desperate to go for a leak, and so on.

2 7
4 7 8 0 9 1 1 6
2 0 3 8 7 2 8

It's like the lottery but considerably less fun and with zero chance of winning a cash prize. Finally, press 'o' if you want to hold for the operator... although you should be warned that she has just gone on two months' maternity leave.

Call centres: The facts

❶ **Your options will be rattled off at bewildering speed.** Prepare yourself mentally. The entire exercise is rather like one of those games where someone brings in fifteen items on a tray, shows it to you for thirty seconds, and then you have to remember as many of the items as possible. And you always forget the thimble.

❷ **Time is money.** If you can't remember which option relates to your query, take a stab in the dark. The calls probably all get put through to the same spotty teenager in any case.

❸ Don't count your chickens. Selecting one option merely takes you to the next level. Say you have pressed '3' about your water bill, another automated voice will then tell you to press '1' if you pay by direct debit, press '2' if you pay by standing order, or press '3' if you pay by cheque. A particularly vindictive call centre may have further steps relating to how much water you consume per week, whether you live in a hard water or a soft water area, and whether you prefer rubber ducks or boats in the bath.

❹ Your call is very important to them. Or so they keep reminding you every time you actually start enjoying the tinny rendition of *Greensleeves* being piped into your ear. If you're call is so important to them, why don't they bloody well answer it?

Call centres: The solution

After being told for the eleventh time how important our call is, most of us swear down the phone and hang up. The patient few who make it through to the bitter end are rewarded with someone who, through lack of training/ education/interest, is predictably of no use whatsoever. And, having failed to resolve your problem, he or she always rounds off by asking cheerfully, 'And is there anything else we can do for you?' 'Yes, but you'd need a proctologist to remove it.'

Here are some practical suggestions:

> Always set aside plenty of time for your call. This is not a task to be undertaken in a hurry. Have suitable provisions handy – sandwiches, thermos, pyjamas, etc – and tell yourself that you're in this for the long haul.

> Put yourself in a calm frame of mind before making the call. Think of distant shores or mountain streams – but not whether or not the water in them is paid for by direct debit.

> Accept that the whole process is fraught with anxiety. You may want to hang up but what if you're near the front of the queue? What if

someone is just about to answer your call?
WHAT IF??

❯ Hum along to the muzak. Not only will it keep
you cheerful, it will help drown it out.

❯ Make sure in advance that you have all the
necessary documentation to hand. Even though
you have been kept waiting on the phone for ages,
the operator is unlikely to do the same while you
rummage through a chest of drawers, wailing
plaintively down the receiver, 'I know it's in here
somewhere.'

❯ Do not follow the example of Teesside's Ashley
Gibbin, who, frustrated at being left hanging on
the phone by NTL's customer services department,
managed to hack into the company's automated
system and erase the pre-recorded message. He
replaced it with his own, which told callers bluntly
that the company wasn't interested in hearing its
customers' complaints. He ended up in court.

Paul English (who is American) is determined to beat call centres and has compiled a database of methods for bypassing the queuing system in order to get through to a human voice.

He says that most companies' systems can be beaten by pressing 'o', 'o *', '* o', 'o #', or '# o' on your phone repeatedly in quick succession. When *Which?* magazine tested the theory, they were able to bypass the systems of npower and Powergen by pressing 'o #' repeatedly. To get through to a human voice at other companies such as NatWest, Alliance & Leicester, and Bank of Scotland, Mr English recommends that callers should ignore all automated instructions and not press any phone key when prompted. A full list of his advice can be found on his website, www. gethuman.com.

Who said there isn't any useful information on the Internet?

EMAIL

It may come as a surprise to learn that email dates back to the 1960s, long before the birth of the Internet. Indeed, in 1976 the Queen proved what a trendsetter she is by sending a minor royal announcement via email.

Email may lack the personal touch of a handwritten letter but it is certainly more convenient than using that old typewriter with the one key that always used to stick. And after a while you'll stop plastering Tipp-ex all over the screen.

Some important email rules

Writing an email is as simple as the average supermodel, but, as with all new-fangled technology, there are rules that need to be followed.

❶ While youngsters might choose streetwise email addresses, we've been around long enough to know that they're the sort of things that can come back to haunt us. A job application is unlikely to be particularly well received if the sender has an email address of boozer@hotmail.com.

❷ Think twice before signing off with an affectionate 'X' or using the smiley face icon (☺) to someone you're sacking by email – it could lead to an industrial tribunal.

❸ Check twice – if not thrice – that the person to whom you're sending your message is the intended recipient, and not the person about whom you are ranting and railing. Email is great for circulating jokes to your friends, but take care that nothing of a risqué nature ends up being sent to someone so timid that the very thought of a bare leg is liable to bring on a hot flush.

Spam

The biggest drawback with email is spam, the electronic equivalent of junk mail. This unsolicited commercial material is forwarded indiscriminately to millions of email addresses every day, leaving elderly spinsters puzzled as to why they should wish for a year's supply of Viagra.

Needless to remark, if you receive an email with any of the following subject lines, it's best to delete it unopened:

Subject: please supply your banking details
Subject: nigerian business proposal very urgent
Subject: YOU ARE A WINNER!!!!!
Subject: Free money!!!!!
Subject: Free university degree!!!!!

In fact, unless you have some very excitable friends, be suspicious of any subject line containing more than one exclamation mark.

Reply all

We've all done it: fired off a well crafted email, or a top-secret one, or a downright spiteful one, only to re-read it and discover with considerable dismay that something's gone horribly, horribly wrong. You've forwarded your private holiday snaps, or the automatic spellcheck has inserted something offensive, or you've moaned about someone and inadvertently sent it to that very person. Surely, if computers are going to take over the world one day, they should have a little more common sense than to let these things happen?

> A helpful HR person at one company sent an employee phone extension list to all employees. But the spreadsheet had hidden columns that were easily unhidden to reveal everyone's pay, bonuses and stock options – including senior management's.

> Government adviser Jo Moore sent an unfortunate email on 11 September 2001, referring to that date as 'a good day to bury bad news'. Somehow the email ended up being splashed on front pages the world over.

> A schoolgirl from Devon mysteriously started receiving top-secret information by email from the Pentagon, having somehow been added to its exclusive email list by a navy commander. One of the emails in question contained advice to the British security forces on how to prevent official secrets from being leaked.

> And who can forget the now-notorious 'Claire Swire email' of 2000, in which the London law-firm employee allegedly emailed Bradley Chait, another lawyer, about an intimate episode they'd enjoyed the previous night – only for the scoundrel to forward it to his mates, who forwarded it to their mates and so on, until three days later it had been read by an estimated one million people all around the world.

It's enough to make you swear off email for good.

STARBUCKS

'One coffee, please.'

'Of course. Latte, Toffee Nut Latte, Americano, Mocha, White Chocolate Mocha, Mocha Valencia, Cinnamon Spice Mocha, Cappuccino, Caramel Macchiato (served upside-down), an iced version of any of the above, Espresso, Espresso Macchiato, Espresso Con Panna, Frappuccino, Mocha Frappuccino, White Mocha Frappuccino, Caramel Frappuccino, Java Chip Frappuccino or Espresso Frappuccino? And would you like that Decaf or Half-Caf; with non-fat milk, low-fat milk, soy milk or organic milk; either breve, dry, wet, extra hot; or with extra espresso shots, extra caramel shots, extra flavour shots..?'

What on earth are they on about?

Starbucks has developed a language of its own – like Greek but marginally less penetrable. To speed communication between employees, the customer's choice is announced in a strict order:

> ❯ whether the drink is iced
> ❯ whether it is decaffeinated
> ❯ the number of shots of espresso (if it differs from the standard recipe for that drink)
> ❯ the size of the cup[2]
> ❯ any added flavourings
> ❯ the type of milk requested
> ❯ any additional customizations
> ❯ finally, the actual name of the beverage.

When you hear the waiter rattle it off, your first thought is, 'Did I really order that?'

---

2 You have a choice of four rather absurd sizes – short (236ml), all (354ml), grande (473ml) and venti (591ml). Although, to confuse matters, in Canada the short is called a piccolo and the tall is known as a mezzo, and you can't get a venti in China.

Oh, for a simple cup of coffee – not something that gurgles away in a manner suggesting that a swamp monster is about to emerge from the cup; not something with so many calories that you dare not eat for another week; and not something that leaves you with so much foam around your mouth that you look like a rabid dog.

How to order an ordinary coffee

The key word to getting an ordinary cup of coffee is 'regular'. To most of us, 'regular' describes someone with enviable bowel movements, but in corporate catering speak it means 'plain'. So if you just ask for a regular coffee – black or white – you should survive the Starbucks experience without the embarrassment of hearing your waiter yell, 'one Iced Decaf Triple Grande Vanilla Non-fat with whip Latte.'

MOBILE PHONES

We have already examined the joys of text messaging (see p.11) but we should not overlook the primary function of a mobile phone, which, naturally enough, is making and receiving phone calls. Back in the days when young Cliff Richard was considered a threat to public morals and binge-drinking among teenage girls meant two halves of shandy, a telephone was a ponderous piece of plastic available in either black or black, and very firmly rooted via various wires and cables to the hallway table. This is no longer the case.

Clunky home phone: Pros and cons

Pros
Never gets lost in your handbag
Makes satisfactory whirring sound
as you dial
Makes you feel like a detective
in an old movie

Cons
Wires get hazardously twisted around you
as you pace back and forth
Repetitive strain injury from dialling
You are not in fact a detective in an old movie

Signalling a new era

The advent of mobile phones means that, instead of queuing outside a call box in the pouring rain while someone armed with a seemingly inexhaustible supply of coins catches up with the previous five years of news from her sister, you can now make or receive a call anywhere in the world.

At least, that is the theory. But for one thing, you can only use your mobile where there is a signal, which, depending on the competence of your provider, could be rather restricted.

One thing you can be sure of is that wherever you really want to use your phone, you won't be able to. What if you're on an underground train that's stuck in a tunnel and you want to tell someone you're running late, or you're in a petrol station kiosk and can't remember your wife's lottery numbers, or you're halfway up Ben Nevis and your map's blown away? Crisis.

In the Devon village of East Prawle, the only signal for miles around is obtained by standing on a park bench on the village green and facing west. Consequently there is a constant queue of people waiting to use the bench.

'I'M ON THE TRAIN!'

Nowadays, the mobile phone is as important an accessory of rail travel as a ticket. But trains are particularly dodgy for mobile phone usage, for three principal reasons:

❶ Your fellow passengers are most likely no longer in awe of the magical mobile phone and will not take kindly to having your inane chatter forced upon them while they're trying to do the *Metro* Sudoku puzzle.

❷ You will invariably have your conversation interrupted by a tannoy announcement from the train manager reminding you for the umpteenth time about the vast range of overpriced refreshments on sale in the buffet or apologizing for the fact that the train is currently running twenty-five minutes late owing to an incident involving a cow outside Market Harborough.

❸ Your call will have an all-too-familiar ending: 'Hi there, Nigel. I'm on the train... ON THE TRAIN! Yes, that's right. I just want to touch base with you on the monthly sales figures. I see we've shifted

650 units at the Tonbridge branch, but only 212 at Sevenoaks and just forty-eight at Ashford. Frankly, I don't think... Hello?... Hello?... Bloody tunnel!... Stupid phone!'

Let's face it, mobile phones are faint at the best of times, the person on the other end invariably sounding as if he or she is talking with a sack over their head. That's why people always shout on mobile phones: it's the only way to be heard.

New-fangled mobile phones: Pros and cons

Pros
You can lie about your whereabouts
You can see who's calling you and choose to ignore them
You can have *The Archers* theme tune as your ringtone

Cons
They show you up as a technophobic old fool
You can't read the screen without the help of
    a small telescope
They have a strange habit of getting lost, misplaced
    or dropped down the loo

Size matters

Size is important with mobile phones. The small ones are light and neat but, by the same token, are therefore easier to lose. Whereas if you have a mobile the size of a brick, it may not look aesthetically pleasing and the weight in your pocket may cause you to lean permanently to one side, but at least there's no danger of mislaying it. For the same reason no one is ever likely to steal it. For years I was the proud possessor of a mobile phone so unwieldy that muggers used to hand it back.

Ring ring

The only advantage to a distinctive ringtone is that it prevents the farcical scenario on a crowded bus where someone's phone rings and everybody on board hurriedly rummages in their bag thinking it's theirs... including the driver.

Otherwise, if you're not someone who likes to stand out in a crowd, stick to the standard ring that comes with the phone. It would have been good enough for Alexander Graham Bell; it should be good enough for you.

FLAT-PACK FURNITURE

Certain phrases fill you with dread: 'Hilarious new comedy starring Jim Carrey', for instance, or 'And now a party political broadcast on behalf of...' Into the same category falls 'We could do with some shelves in the spare bedroom.'

And yet every bank holiday thousands of couples stagger from out-of-town stores wielding a bulky flat-pack, a spirit level and a huge tub of optimism, lulled into a false sense of competence by the promises of the store catalogue, only to end up with kitchen stools so fragile that no one is allowed to sit on them; coffee tables with legs at all angles, like Bambi on ice; and shelves that slope so much that the ornaments have to be glued down.

Promise (Lie)	Truth
Your home will look unique and classy.	Your home will look like Croydon's Ikea showroom.
Your home will look individual.	Your home will look like your neighbour's home.
You and your partner will want to gaze lovingly at the furniture you have assembled together.	You will want to file for divorce – and force your other half to take custody of the wooden monstrosity in your formerly harmonious home.
You and your partner will be transformed into grinning, young, blond Swedish types with the perfect home to match.	You will vow never to attempt DIY again, as you sit amongst the ruins of your shelves, sweating and swearing in equal measure.

DIY: The inevitable disasters

Once you have penetrated the pack, you find yourself immersed in a mysterious world of dowels and flanges, and the ever-present small tube of glue. The glue seems to be included in the hope that if nails and screws don't hold the contraption together, some runny Pritt-Stick should do the trick.

Nevertheless, you will be full of enthusiasm. STOP! There are some inevitable annoyances you will have to overcome:

> You have bought the AJ812 but rest assured the instruction manual will be for the AJ813. Similar, but not the same; in fact, just enough subtle differences to turn the perfect design on the pack into something that looks as if it has been assembled by Frank Spencer.

> You will end up opening the illogically rip-proof packaging with your teeth. Dentists must be rubbing their hands with glee at every flat-pack purchase.

> There will be one small – but very vital – pin missing. Without it the contraption cannot stay together in one piece.

> You will, at some point, feel like smashing the entire thing to bits with your hammer.

Universal step-by-step DIY instructions

With everything in place, it is time to consult the easy-to-follow, step-by-step guide, which usually looks as if it's been printed on a scrap of paper by a child with a John Bull Printing Set. If you cannot make head nor tail of the diagrams, try turning them upside down.

Instruction: Fit dowel pin A into plug B.
Reality: Dowel pin A, once located beneath side panel
   2, is wider than plug B. Either the plug hole must
   be widened or the dowel pin must be sanded. This
   procedure takes another half-hour and is repeated
   throughout the assembly whenever something must
   be fitted into something else.

Instruction: Ensure side panel 1 sits flush with end panel Q, and fix them using nail pin 4.

Reality: The gap between side panel 1 and end panel Q is wide enough to drive a toy car through, let alone nail pin 4 – which, incidentally, is missing from the pack.

Instruction: Now stand the chest upright and insert drawers A to D onto runners 3, 5, K and X.

Reality: Drawers? I thought this was supposed to be a chair.

If Adam and Eve had been made by flat-pack furniture manufacturers, there would be no human race.

DIGITAL CAMERAS

Just as opening an Easter egg to find that two-thirds of the box is packaging, so the worst thing about opening a digital camera box is discovering that the vast bulk of it is taken up by the handbook.

The general rule of thumb seems to be that the manual shall be at least twice the size of the camera, forming an epic publication to rival *War and Peace*. It is a truly daunting prospect, invoking all manner of terrifying-sounding things such as 'white balance' and 'ISO-based image stabilization', particularly when you recall that the instructions for your old Kodak Brownie amounted to little more than two pages of diagrams.

How to work a digital camera

❶ Chuck the manual in the bin.
❷ Turn the camera on. This is quite important.
❸ Choose the 'auto' setting, whereby the camera makes the crucial decisions about lighting, focus and distance. Millions of pounds have been spent developing the camera's technology, so the chances are it knows more than you do.
❹ Point.
❺ Click.

Don't be intimidated

Unless you're planning to shoot a remake of *Ben-Hur* in 'movie' mode or to mount an exhibition at the National Gallery of your holiday pictures from Morecambe, you don't need to know about most of the camera's functions. David Bailey may be familiar with his focal range in 'super macro' mode, but as long as your thumb isn't over the lens and your hand isn't shaking like a cocktail barman's, you'll be fine. Digital cameras are virtually foolproof, which, let's face it, is just as well.

CASH DISPENSERS AND OTHER AUTOMATIC MACHINES

When we wanted some money back in the day, we used to go along to our local bank, wave our chequebook, have a pleasant chat with the friendly cashier about the weather, and emerge with our £10. It was quite the social transaction. But now, in the name of speed and efficiency, human beings have been replaced by machines.

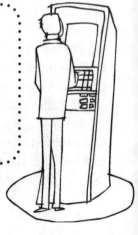

Not many people know – or indeed care – that the world's first cash dispenser was opened by Reg Varney of *On the Buses* 'fame' at Barclays Bank, Enfield, Middlesex, on 27 June 1967, and that the maximum withdrawal at that time was £10.

ATM etiquette

> Never make smalltalk with people in the queue. Everyone's trying to ignore the fact that they're about to conduct their personal banking in the company of strangers, right in the middle of the high street.

> Rather like men standing side by side at a urinal, never, NEVER let your gaze drift as the person beside you taps in their PIN.

> Be watchful of anyone suspicious standing nearby, particularly if they are wearing a balaclava.

> Don't say, 'Gosh, I'm rolling in it!' when checking your balance.

> Don't say, 'Gosh, you're rolling in it!' to the stranger in front of you.

ATM lingo

Advice slip: Contains no advice whatsoever, merely a print-out of your bank balance – which is more likely to alarm than advise.

All other services: Alas, no. You cannot request marriage counselling or a portion of chips. You may change your PIN and that's pretty much it.

Type in desired amount: As long as you desire a multiple of £20.

Your account has insufficient funds: A polite way of saying 'Forget it: you ain't got no cash.'

ATM pros and cons

Pros	Cons
Open 24 hours a day.	Increased likelihood that you'll decide to withdraw £200 after a night at the pub and treat your friends to a black cab ride around the city.
They're on every street corner.	As are muggers, drug dealers and other ne'er-do-wells with designs on your cash.
No need for inane chit-chat with a cashier.	Increased need for inane chit-chat with the tramp sitting beside the machine.

Other dreadful machines

The bank, the cinema, the train station, the airport, the library, the car park and even the supermarket have introduced automated machines to conduct trans- actions. And make no mistake: these machines can be scary, devouring the unwary like a Venus fly-trap. Some claim to be able to accept notes, but no matter how you angle the Queen's head, they always seem to spit them out.

If a machine of any sort seems not to respond to your furious button-pressing, received wisdom suggests the following reactions might help:

> Hit it
> Kick it
> Scream at it
> Plead with it
> Swear and storm off, muttering put-downs within earshot of the machine.

Self-'service'

There is no machine in this world more capable of turning even the most techno-loving person into a violent Luddite than the supermarket self-service till. What lunatic came up with this thing?

> ❯ It takes at least three times longer than going to an old-fashioned till, even if said till is manned by one of those chatty old ladies.

> ❯ The stupid things keep alleging that 'there is an unexpected item in the bagging area' and going on strike. 'THERE IS NOTHING IN THE BLOODY BAGGING AREA!' you shout at the faceless screen, smashing your shopping bag up and down on the offending shelf.

> ❯ Most items need to be authorized by a member of staff in any case (the fun ones anyway, e.g. wine).

To give them their due, these machines have been programmed to speak in a flat, monotone voice so that customers will still feel they are being served by real cashiers. As the technology improves, we can expect a machine that files its nails, talks to other machines about last night's date, and is programmed to scream, 'How much are the large tubes of haemorrhoid cream, Tracey?'

GLOBAL WARMING

With the possible exception of Simon Cowell, global warming is the biggest threat to this planet. Who says so? The Experts, so it must be true, even though they're the same Experts who, after a couple of particularly cold winters in the late 1970s, proclaimed that we were about to enter a new Ice Age. Now it seems that nobody is keen to place a bet regarding the prospect of penguins being seen swimming under Tower Bridge or polar bears stalking seals in the centre of Walsall.

Is it all an unseasonally tropical storm in a teacup?

A few years ago, TV gardening experts advised us to put Mediterranean plants in our garden to combat the summer droughts that were predicted as part of global warming. Most of these plants have since rotted under summer downpours.

And on the very day in March 2007 that the experts were declaring that the unseasonably high temperatures were yet another indication of global warming, the newspapers reminded us that a year earlier parts of Kent were under several inches of snow.

Those of us who have been around for a while know that if you had to pick one word to sum up Britain's weather, it would be 'changeable'. That's why we talk about it all the time.

The joys of global warming

It's not all doom and gloom. Why not embrace global warming?

> Sunbathing in February
> Skiing in May
> Less call for itchy knitted sweaters
> More call for evening drinks al fresco
> Plenty of interesting weather events to make smalltalk about in the Post Office queue
> No need for disappointment at the lack of snow at Christmas – simply forgo the anticipation: it's just not going to happen ever again
> Palm trees and parakeets in your herbaceous border

Conspiracy theories aside...

I don't want to be held personally responsible for destroying the planet. What a thing to have to live with... or not. So here are a few tips to help reduce your carbon footprint from a size 11 to a size 7½:

> You never see politicians or members of the Royal Family flying anywhere, so follow their lead. Next summer, instead of catching a plane to go abroad, choose a holiday destination within easy walking distance of your home. Camping in your neighbours' front garden for two weeks is just one of many fun-packed options.

> Embrace alternative methods of travel. Sell the 4x4 and commute to the office by greener modes of transport, such as the pogo stick, roller skates or Spacehopper.

> Swap your Cadillac for a G-Wiz electric car. It may appear small but it can readily accommodate four passengers, provided three are clinging to the roof. And while it is not terribly robust (it would probably come off second best in a collision with a bicycle) and has the acceleration of an asthmatic man climbing a steep hill, it is eco-friendly.

> If you must drive your car, travel light to use less fuel. Hauling an extra hundred pounds in your vehicle reduces your fuel economy by up to two per cent, so before setting off on a journey, always leave your family behind. If you happen to forget, don't be afraid to push them out en route.

> Cut down on electricity by removing all light bulbs in your house. Use candles instead and paste fluorescent panels on to the walls to guide you from room to room. You could try breeding fireflies and glow-worms for additional natural energy. Encourage your family to wear bright clothes so that you can locate them in the dark. Paint your hair (if applicable) lime green.

> Keep the heating in your house to a minimum. Wear several layers of clothing and, accordingly, widen doorways to enable you to get around the house. Ask the council about the possibilities of having your spouse lagged for the winter.

> Don't waste water and electricity by washing and blow-drying your hair unnecessarily. Try wearing it in dreadlocks instead. It could grow on you.

> Install low-flow showerheads in order to use less hot water. Never take a bath. Ever again. Don't worry if you stink to high heaven. That way, everyone around you will soon be going green, too.

FACEBOOK, FRIENDS REUNITED AND OTHER SUCH THINGS

'Your school days are the best days of your life.'

No wonder so many teenagers suffer from depression: if spots, double maths, endless exams and that French exchange where you had to share a room with the weird little brother are as good as it gets, what on earth is there to look forward to in the future?

$$\sin^2 \alpha + \cos^2 \alpha = 1$$

$$\int_0^x \frac{t^n dt}{e^t - 1}$$

$$x = \frac{-b \pm \sqrt{b^2 - 4ac}}{2a}$$

$$ax^2 + bx + c = 0$$

$$(a - b) = a^2 - b^2$$

$$tg\alpha = \frac{\sin \alpha}{\cos \alpha}$$

The old lie

With hindsight, most of us would probably agree that schooldays were *not* the happiest days of our lives. You worked hard – for no money – at subjects you didn't particularly enjoy. Algebra, Wordsworth, the Agrarian Revolution, French irregular verbs, valency tables... the list was endless. As a break from study, you were subjected to mental and physical torture out on the perpetually drizzly sports field.

Yet in spite of all this, in moments of idle thought, we often hark back to our schooldays and wonder whatever became of the girl who chained herself to the school railings on the last day of term and refused to free herself until Donny Osmond was brought to her? Did she ever get to meet Donny? Is she still chained to the railings? Or what about the boy who used to stick peanuts up his nose and once stole the geography teacher's car? Is he in Wormwood Scrubs or Strangeways? Enter the new-fangled craze of social networking.

The first five things you will do on Facebook

❶ Look up all your exes. Don't deny it.

❷ Look up all your colleagues and feel rather like a stalker when you stumble across their holiday photos.

❸ Upload the least old-fogeyish photo of yourself to show any snooping exes and colleagues how young and fabulous you still are.

❹ Upload photos of you on holiday with your other half to show any snooping exes and colleagues how popular and dynamic you still are.

❺ Log out and then log in again ten minutes later to see if anyone's friend-requested, poked, messaged or tagged you. (These words will be part of your vocabulary by this stage.)

What could possibly go wrong?

> Nobody accepts your friend request except that horrible girl with pigtails who is still demanding to know what happened to her pencil case thirty-odd years ago.

> You are bombarded with messages from an old flame you eventually managed to lose touch with after a tortuous break-up, pleading with you to explain how it all went wrong.

> You will meet up with someone at whom you once stared longingly across the playground, only to find them a bitter disappointment in their dotage. Likewise, you'll discover that the school bully has actually done rather well for himself, shattering your happy delusion that he'd amounted to nothing, as your headmistress always predicted.

> 'The Curse of Friends Reunited' may strike. The online rekindling of old passions has been responsible for the break-up of more marriages than Zsa Zsa Gabor.

Social networking lingo

Some people are disconcertingly honest, admitting to being 'a disaster in relationships' or 'emotionally unstable', but all sane people airbrush anything too awkward or embarrassing from their recent past.

Claim	Truth
'I am an exhibited artist'	Is a painter-decorator
'I work with some big names'	Sells books
'I was a professional footballer'	Had unsuccessful trial with Grantham Town
'I'm a major player in the UK fashion industry'	Has a loyalty card at Dorothy Perkins
'I'm big in international shipping'	Manages the gift shop on the cross-Channel ferry
'I rode the dotcom boom'	Learnt how to use the Internet

Stick to the truth – or at least a slightly sanitized version of it, omitting any gory bits. These are your old school acquaintances – they don't want to hear about your assorted ailments or the full breakdown of your breakdown after the breakdown of your second marriage.

One avenue of pleasure on Friends Reunited that had to be suspended was the bulletin board, after the teachers' union complained that its members were being subjected to unfair comments. In fact, a Lincolnshire man had to pay £1250 damages for libelling one of his old teachers on the website.

So if you're looking for an outlet to redress any grudges, forget it. The history teacher who was caught with the sixth-former behind the bike shed, the maths master who could hurl a board rubber the way Steve Backley used to hurl a javelin, and the English mistress with severe personal hygiene problems: they're all immune.

Facebook and Friends Reunited have hundreds of millions of registered users. If you are curious as to what happened to people from your past, perhaps the best advice is to scour the site and, when you have satisfied yourself that nobody you went to school with has done better than you, quietly return to your life.

HI-TECH TV SETS

It has to be said that comparing the picture quality on an old black-and-white with today's colour marvels is like comparing Jackie Collins to Shakespeare. But I'll do it anyway.

Definition of a TV, back in the day

> Aesthetically displeasing wooden thing with a thick, nine-inch glass screen that sat in the corner of the living room between the wireless and the gramophone.

> Everybody, from the craggiest newsreader to the most glamorous star, appeared in varying shades of grey.

> There were nearly always vertical lines on the screen. In between attacks of vertical lines, the picture would slip slightly, the result being similar to a hall of mirrors where people have abnormally large heads and short bodies.

> The picture would be obliterated in a manic mass of squiggly lines if anyone within a ten-mile radius was using an electrical appliance. You could ruin the Queen's Speech for the entire neighbourhood just by vacuuming the crumbs from Christmas dinner.

Definition of a TV, nowadays

> Widescreen

> Flatscreen

> Splitscreen

> Hugescreen. Every time you sit down to watch *Coronation Street* you expect an usherette to appear with a tray of King Cones and choc-ices.

Plenty of knobs

Old TVs had a range of knobs to twiddle at the front of the set – horizontal hold, vertical hold, contrast, brightness, and volume. We used to play with these until something resembling an acceptable picture returned. Indeed, careful turning of the vertical hold knob created a form of split-screen technology years ahead of its time.

If the problem was unable to be solved by knob-twiddling, the TV repairman was called out. He would study the set, twiddle the knobs, take the back off and put the back on, before announcing solemnly: 'It's yer valve.' Whatever was wrong with any old black-and-white set, it was *always* the valve.

How to fix a modern TV

Don't bother: you can't. For all their streamlined elegance, cinematic sound and pin-sharp picture in glorious techni-colour, there are no knobs to twiddle. Instead you have to call out an engineer who will then take the set away for a few days, tell you it was a much bigger job than he had anticipated and charge you accordingly. If you ask him what was wrong with it, he'll mutter something about having had to solder the component parts because the picture frame circuit had failed. In old money, it was the horizontal hold.

Expecting the unexpected

Sales staff working on commission will go to great lengths to persuade you to take out a warranty, offering you anything from a free pen to their sister's hand in marriage in return for three-year cover. But, as with any form of insurance, warranties don't cover everything. Look through your home contents insurance policy and you will see that, while your garden fence may be fully covered against being trampled down by a herd of rampaging llama or being wrecked by a meteorite, it is not insured against a gust of wind. Anything that is likely to happen is conspicuous by its absence.

> **Although the hi-tech gadgetry of a thirty-six-inch plasma, low-fat, boot-leg, widescreen set with catalytic converter may sound terribly impressive, the only question you really need to ask is: will it get *EastEnders*?**

CRIME

Naturally crime is the fault of the younger generation. They're so mindless. They listen to gangsta rap and they want to become gangsters. We listen to rock music but it doesn't want to make us want to smash up the local Travelodge. The only letters many of today's kids are going to get after their names are ASBO, yet their parents or parent still defend them to the hilt and even appear to have devised a series of accepted euphemisms to describe their offsprings' behaviour:

> 'He's no angel' means he's a dangerous little psychopath.
> 'Cheeky' means she spits at passers-by and once attacked her granny with an iron bar.
> 'Lively' means out of control.
> 'He's just a typical teenager' means he rules the neighbourhood by fear.

Locating a police officer

Nowadays it's as hard to find a police officer on the beat as it is to spot a small nun in a penguin colony. But given that it took four officers to arrest Chris Tarrant on suspicion of throwing cutlery and eight to interrogate Jose Mourinho's dog, they are presumably out there somewhere. You just have to know where to find them.

Is it a crime, officer?

Crime is on the rise, although the government and the police are always able to spin the statistics to make it appear as if they are on top of the situation. 'It is true,' they say, 'that incidents of robbery, burglary, street crime, violent affray, car-theft and murder are on the increase, but look at the positives: incidents of impersonating a Chelsea Pensioner have decreased markedly.' That's OK, then.

So before you report a crime – or indeed commit one – it's best to check that it is in fact a criminal offence. (Impersonating a Chelsea Pensioner really is.)

Not a criminal offence:

> Being a topless woman in Liverpool – provided that you work in a tropical-fish shop.
> Shooting a Welshman in Hereford – provided that you are in the Cathedral Close on a Sunday and use a longbow.
> Shooting a Scotsman in York – provided that he is carrying a bow and arrow.

> Shooting a Welshman in Chester – provided that you are within the city walls after midnight and use a bow and arrow.
> Relieving yourself anywhere you like, including in a policeman's helmet – provided you are a pregnant woman.

A criminal offence:

> Being drunk in a pub.
> Transporting corpses or rabid dogs in your taxi, or hailing a taxi whilst suffering from the bubonic plague.
> Entering the Houses of Parliament in a suit of armour.

> Dying in the Houses of Parliament.
> Urinating in public, unless you do so on the rear wheel of your car, with your right hand resting on the vehicle. (Or are pregnant – see above.)
> Eating mince pies on 25 December.

Reporting a crime

You could always try phoning the police, but you'll probably be put through to a call centre on the other side of the world, and in any case, rather like doctors, the police no longer do home visits if they can help it. So the personal touch may reap greater reward – although your local police station has almost certainly been closed down, while the one in the next town has been turned into an All Bar One.

Once inside the station, there is a right way and a wrong way of persuading the police to take an interest in what you have to say. If you get it right, the desk sergeant will whisk you straight off to an interview room; if you get it wrong she'll go back to reading *Heat* magazine.

WRONG: 'I've just been involved in a bad car crash. I think everyone's OK but it's a terrible mess out there. There's glass and metal everywhere.'
RIGHT: 'I'm afraid I've just reversed into one of your patrol cars.'

WRONG: 'A gang of kids has been terrorizing us for weeks. They've smashed all our windows and we're afraid to leave the house.'
RIGHT: 'I caught one of the gang the other night and gave him a clip round the ear.'

WRONG: 'Quick! A woman's being held hostage in our street. He's got a knife.'

RIGHT: 'A woman's being held hostage in our street. He's got a knife. I think she's a friend of the Chief Constable.'

And so on.

BOOKING HOLIDAYS ONLINE

Booking a holiday is something of a hit-and-miss affair. Unless you are returning to Mrs McGonagle's guesthouse in Prestatyn for the seventeenth consecutive year, you are largely reliant on the honesty of brochures and, in these flashy techno times, websites. At least nowadays you are likely to get a recent-ish photo of the resort (with cranes and football fans conveniently airbrushed out); back in the day you'd get nothing more than line drawings, which, in terms of their accuracy, might just as well have been done by Picasso.

Holiday lingo

Never mind mastering Spanish or Greek; your main challenge will be translating the holiday brochure or website.

'Five minutes from the sea'	By Concorde
'A hundred metres to nearest restaurants'	One greasy-spoon caff down the road
'Secluded beach'	Impossible to reach other than by parachute or yak
'Popular beach'	Can't see the sand for people
'Shingle beach'	Like the surface of the moon, but slightly less picturesque
'Enjoys pleasant sea breezes'	Bring your thermals
'International cuisine'	Pizza
'Nightly entertainment'	A pack of cards in the lounge
'Partial sea view'	If you lean precariously over the balcony
'Lively atmosphere'	Popular with stag parties
'Friendly staff'	Lock up your daughters

Booking online

Booking your flight online is straightforward and can be done in just a few minutes, but for goodness' sake take heed of a few simple lessons before entering your credit card details:

❶ Geography. When selecting your airport of departure, make sure that, if you're planning to fly from Birmingham, England, you don't accidentally click on Birmingham, Alabama. It's a long walk home.

❷ Economics. Flight prices with budget airlines can fluctuate wildly, not only from day to day but sometimes from hour to hour. If you're not in a hurry to book your holiday, visit the site for a few days beforehand to see if the price changes. Contrary to popular belief, it does not necessarily follow that flights booked months in advance or at the last minute will be any cheaper than those in between. When it comes to airline fares, logic flies out of the window.

❸ **Maths**. The figure you are first quoted is by no means the final price. The thought of flying from London to Barcelona for 99p may seem enticing but by the time baggage fee, insurance fee, check-in fee and a donation to the pilot's animal charity of choice have been added on, the price has rocketed to nearer £100. The most irritating 'add-on' of all is the credit or debit card fee. There's no other way to pay! Try writing a cheque to a computer or stuffing ten-pound notes in the hard drive...

International dictionary

There are some scenarios that crop up time and time again in foreign climes. Learn these phrases in the native tongue of the country you are visiting; they will prove invaluable.

'No thanks – I do not wish to visit your friend's ceramics shop.'

'We have plenty of carpets at home, thank you.'

'Could I have chips instead of rice/ vegetables/ this bowl of unidentifiable matter?'

'Gosh – salami/ cake/ curry for breakfast. What a novel idea. Do you perchance have porridge?'

'Which way to the pharmacy? I seem to have a dodgy stomach.'

WASHING MACHINES

How much simpler life was when washday consisted of a scrubbing brush, a washboard and a mangle. It may have been crude and it may have been hard work, but at least there wasn't much that could go wrong, apart from contracting the odd case of scrubbers' elbow.

By contrast, today's automatic washing machines have more knobs and dials on them than a NASA spacecraft. They're a callout waiting to happen. To group them under the heading 'domestic appliance' is grossly misleading because that implies they are tame and housetrained. Nothing could be further from the truth.

Where to start?

If you have bought a new washing machine, just looking at the front of it is enough to make you giddy. There are programmes for cottons, synthetics, delicates, wool, hand wash, cold wash, rinse, drain and spin; temperature choices of 90°C, 60°C, 50°C, 40°C and 30°C; spin speeds of 1200, 900, 700 and 500; not to mention optional extras like start delay, prewash, intensive rinse, rinse hold, spin, and quickwash. And that's just on a basic model. There are fewer programmes on Channel 5!

❶ Ignore all washing machine and detergent instructions; they wilfully contradict each other.

❷ Put the detergent in the drum. If you put it in that top drawer it always leaks all over the floor. These things were sent to test us.

❸ Choose a cycle. Conventional wisdom suggests choosing cycle 'D' (which stands for 'Don't know'), but if you are in any doubt whatsoever, simply choose the coldest option. Easy peasy.

As a rough guide, any of the following may indicate that all is not well with your machine: smoke, flames, wailing sirens, a pool of water the size of Lake Titicaca. If any of those occur, it could be time to fetch that old washboard out of the garage.

IPODS

It wasn't so long ago that Sony Walkmans and Filofaxes were the cutting edge of technology. Everybody in the 1980s absolutely had to have them, although the less trendy were perfectly happy with a transistor radio and a diary. Now it's all digital, which just seems a good excuse for getting rid of all the old stuff that worked perfectly well and forcing people to spend money on the latest innovation.

iWhat?

An iPod is actually an MP3 player, which in turn is another name for a digital audio player. Are you still there? It just happens that the iPod is currently the bestselling line, but let's not overwhelm ourselves with too many possibilities.

iPod pros and cons

Pros	Cons
You can store 20,000 songs	Most of us couldn't name 200 songs
You can set it to random selection and be unaware of what's coming next	You could switch on the radio for free
You are sealed off from the outside world	There are people with iPods who don't even know there's a war on

The iPod cynic

Frankly, it's not as if the music on today's iPod sounds any better than it did on the old Walkman when you hear it across the train carriage. To the distant listener, it still produces that strange, indistinct, hissing sound, reminiscent of a jazz trio where the drummer is playing the cymbals with pastry brushes. No matter how hard you listen, you can never quite work out what the song is. What sort of invention makes Karen Carpenter sound the same as Snoopy Dogg or whatever it is he calls himself?

If you're using an iPod on public transport, not only can you not eavesdrop on other people's conversations – which, let's face it, is usually the most interesting thing to do on a long train journey – but you might be missing something really important, like your stop. Instead of getting off at York, you could end up in Doncaster – and surely nobody deserves that. At our age, we need all our faculties in full operational mode; it's no use having one of the major senses on standby.

iPod dictionary

Rip	Copy
Burn	Copy
Upload	Copy
Download	Copy

Old-style downloading

Of course, we were downloading music decades before anyone had ever heard of an iPod, except we didn't call it 'downloading', we called it 'taping'.

Joyously we used to tape songs from the radio or from records onto cassettes, only to hear our favourite track ruined by the rumbling sound of a passing train or by someone shouting in the background. I have often felt that George Martin missed a trick by not asking somebody's mum to shout 'Your dinner's ready!' during the chorus of *Hey Jude*. It definitely added something to the piece.

Acceptable iPod behaviour

If you do decide to invest in an iPod and sit with the hoodies at the back of the bus, it is important to try to look cool.

You must not:

> Sing along – particularly if you're listening to Lionel Richie or anyone of his ilk
> Dance – under any circumstances
> Headbang
> Play air guitar – or, worse, if you're listening to classical music, air violin or air harp

You may:

> Nod your head subtly and rhythmically from time to time. You should do this even if you haven't managed to switch the iPod on. Nothing is worse than having to admit in public that you don't know how to work the latest gadgetry.

TELEPHONE ANSWERING
MACHINE MESSAGES

We either lead such hectic lives that we're never around to answer the phone in person – or we want people to believe that we lead hectic lives. Thanks to the invention of the telephone answering machine, the caller doesn't know whether you're sitting there quietly playing solitaire, waiting for the clock to reach an hour when it's not too embarrassingly early to go to bed, or whether you're out painting the town red... or at least a dark shade of beige.

But in order to enjoy the mystery that your perceived absence will create, it is necessary to record a message on the answering machine. Most machines offer their own pre-recorded outgoing messages. They're usually spoken by a woman from Tunbridge Wells, so unless you happen to be that woman from Tunbridge Wells, it is best to record your own message.

Getting started – some rules

❶ **Recording your own message is not a task to be undertaken lightly.** It should be treated with the gravity of an audition for the Royal Shakespeare Company. First, you need to write your script, although neither Hamlet's soliloquy nor King Lear's ravings on the storm-swept heath are particularly suitable for telephone answering machine messages.

❷ **Your message should be short, simple and to the point.** If you ramble on for the full two-minute allowance, the caller will lose the will to live.

❸ **Read the instructions.** Unless you have chosen a £3000 diamond-encrusted model, these will not, for once, run to a second volume. And there is nothing more unprofessional-sounding than an answering message that ends in '...Christ, how do I switch this bloody thing off?' followed by eighty seconds of swearing and white noise.

Phrases best avoided

'I'm on holiday at the moment but will be back on the seventeenth.'
Burglars will see this as an open invitation.

'I can't get to the phone' or 'I'm a bit tied up at the moment.'
A caller with an over-active imagination will notify the emergency services.

'I am currently on the loo and unable to get to the phone.'
Not even the closest of friends needs this level of detail.

Steer clear of saying anything you may later regret. A message along the lines of 'Sorry I can't take your call because I'm too busy counting all the money I'm making for doing very little at Whiting, Halibut & Chubb' might be amusing... until Mr Halibut calls you unexpectedly at home.

Be prepared

> **Nobody likes the sound of their own voice. Accept this.**
> Just speak clearly and brightly, but don't put on airs and graces. Think 'toothpaste ad' rather than 'Queen's Christmas speech' and you'll be fine.

> **Leave the jokes out.**
> Unless you are a professional comedian with a reputation to uphold, don't try to be funny. You may do the best 'only gay in the village' impression for miles around but it could puzzle your local library when they call to tell you that the archaeology books you reserved are ready for collection.

> **Don't rush.**
> You'll only wear yourself out – and the sound of heavy breathing is particularly off-putting on a phone line.

MODERN TERMINOLOGY, INCLUDING POLITICAL CORRECTNESS

If you get a group of middle management types together, two thoughts immediately occur to you:
   a) why is there never a sniper around when you need one? and
   b) what on earth are they talking about?

Middle managers are to business what fleas are to a dog: they serve absolutely no useful purpose but are incredibly irritating and seem to be everywhere.

All too aware that their ridiculously elevated position and its attendant lifestyle are under constant threat and that they will eventually be found out, they seek to conceal their inadequacies beneath a veil of gobbledygook. They figure that if they make themselves sound important, someone somewhere will believe that they really are. It is a sad indictment of modern companies that the tactic invariably works.

Jargon dictionary

In a world where your significant other may be a mover and shaker in human resources but is liable to be a victim of downsizing, restructuring, re-evaluation, rationalization, streamlining or a reshuffle at any time, here are some examples and explanations of management speak:

Sing from the same hymn sheet, as in 'I need to make sure we're singing from the same hymn sheet, Piers.' All it means is, 'I need to make sure we're in agreement.' So why don't they just say that?

Think outside the box, as in 'I reckon we need to think outside the box on this one.' Which box? Roughly translated, it means 'We need to look at it from a new perspective.'

Game plan, as in 'What's your game plan on the chocolate sprinklies project?' 'Game plan' is simply another term for 'strategy'. But its only rightful place in the English language is when referring to sports tactics. Alex Ferguson might have a game plan; a guy trying to sell chocolate sprinklies does not.

Push the envelope, as in 'This deal is so important to us – we need to be pushing the envelope.' It means pushing something to the limit, going beyond the commonly accepted boundaries, being innovative... all the things middle management people generally are not.

Circle back, as in 'Hope you don't mind, but I'm just circling back to you on this.' It means 'I'd like a progress report on the task I only gave you yesterday afternoon, and which I know will take a week to complete, but it makes me look on the ball, as if I know what I'm doing.'

Multi-tasking simply means doing more than one thing at the same time. Housewives have been multi-tasking for decades. For middle management, singing while you're on the loo constitutes multi-tasking.

No-brainer is a term that should be applied to whoever uses it. In fact, it means something that is obvious.

Bring to the table, as in 'I didn't think Piers brought much to the table today.' This has nothing to do with Piers forgetting the crisps and the prawn vol-au-vents; it means he didn't contribute much to the meeting and was of little value.

Brainstorm means nothing more than discuss. Misleadingly it implies the presence of intellect.

Touch base, as in 'Let's touch base on this tomorrow morning.' It just means 'Let's talk about it tomorrow morning.' Most employees are wary of touching anyone's base these days, lest it lead to a charge of sexual harassment.

Team leader is the nominated head of a group. But the image conjured up by the term is of huskies pulling an Antarctic expedition, not of a bunch of suits trying to look impressive at a paperclip conference.

Fast-tracking is a synonym for taking a short cut. The CEO's son-in-law stands a good chance of being fast-tracked through the company.

Radar, as in 'I'd like to get on your radar for a meeting next week.' What tosh! All it means is, 'Can we arrange a meeting for next week?' No doubt with a bottle of equally pretentious sparkling water for refreshment.

Thought leader refers to either a person or a company seen as being at the forefront of a particular field. Thus John George Haigh could be said to have been a thought leader in the field of murdering people and disposing of the bodies in baths of acid.

I hear what you say is a particularly patronizing and dismissive phrase. It is always followed by 'but...'

Guesstimate is an even rougher estimate than an estimate. It is an example of a portmanteau – a new word formed from two old ones, such as 'affluenza' (a contagious condition of debt, anxiety, etc caused by the dogged pursuit of wealth); 'brunch' (a late morning meal that combines breakfast and lunch); and 'infotainment' (a media broadcast that combines hard news and celebrity interviews, i.e. it is a mixture of information and entertainment). But like so much of today's vocabulary really it's all 'strattle' (a combination of stupid and prattle).

Run it up the flagpole and see if anyone salutes means to present an idea and see what reaction it receives. Many middle managers should be run up the flagpole and left there.

Adding bandwidth simply means hiring new staff. Since when have people been reduced to computer measurements? And don't you hate it when someone says 15k instead of fifteen thousand?

Mindset refers to how a person thinks or behaves. Thus Roger Federer has a straightset mindset, Casey Jones had a trainset mindset and Barbara Cartland had a twinset mindset. On the other hand, Jade Goody's mind is set permanently at zero.

There is no 'I' in 'team' is a cliché trotted out to emphasize the importance of teamwork over individuality. But there is an 'M' in 'team', and an 'E'. So how does that work?

Mission-critical, as in 'Delivery of this document on time is mission-critical.' This absurd twaddle means that delivery of the document is crucial to the successful completion of the project. You can guarantee that nobody saying it has ever been anywhere near a space shuttle.

Proactive means taking responsibility, as opposed to just sitting around idly doing nothing. Few middle managers are proactive, though all like to think they are.

Win-win, as in 'I love it, Piers. It's a win-win situation.' This is a strategy where nobody apparently loses, although it is often little more than corporate spin, since the reality is that it's tantamount to a humiliating, face-saving operation. Although he may not have realized it at the time, the Big Bad Wolf was in a win-win situation when he tried to blow down the Three Little Pigs' houses. If he had succeeded, he would have had pork for dinner for the rest of the month; but even by failing he was immortalized in folklore, thus guaranteeing his surviving family healthy profits from book sales, audio and video rights, use of image, etc.

Political correctness

Meanwhile political correctness has taken over the world. And a fat (sorry, gravitationally challenged) lot of use it is! Everyone seems to talk in code.

English	PC English
drunk	an anti-sobriety activist
bald	comb-free
a dish-washer	a utensil-sanitizing executive
dishonest	ethically disoriented
lazy	motivationally challenged
unemployed	involuntarily leisured
old	gerontologically advanced
poor	economically marginalized
homeless	residentially flexible
a housewife	domestic engineer
beer gut	a liquid grain storage facility
a psychopath	socially misaligned
a bad hair day	rebellious follicle syndrome
ignorant	intellectually unencumbered
shy	conversationally selective
a bad cook	microwave compatible
a big nose	nasally gifted
short	anatomically compact
clumsy	uniquely co-ordinated
dead	living impaired

How long before we have to rewrite history so that William the Conqueror becomes William the Kingdom Oppressor, Ivan the Terrible becomes Ivan the Niceness Deprived, and the Virgin Queen becomes the Monarch Who Was Not a Previously Enjoyed Companion?

DIGITAL CLOCKS AND VIDEO RECORDERS

It's a funny old world. We can remember the intricacies of the local train timetable but we can never remember whether the clocks go forward or back in March. And even if we do get it right, we can never remember what it means: will it be lighter or darker in the mornings? The favoured method of remembering the rule is:

Fall (autumn) backward, Spring forward

But as a memory guide that is by no means perfect. Drunks are just as likely to fall forward as backward, and if you startle a cat, it springs backward. And Zebedee off *The Magic Roundabout* merely seemed to spring straight up and down.

Anyway. Adjusting a digital clock for Daylight Saving is a considerable challenge. No two clocks are alike, it seems. They all have a button called 'clock', but beyond that you're on your own. It's less a case of scientific application and more a case of trial and error.

Mastering a digital clock – in theory

You press the 'time' button and the hour advances by one. If this happens, give yourself a pat on the back, for it appears that you have somehow unlocked the secret of the clock's timing mechanism with the minimum of disruption. All you have to do is repeat the trick with the minute numbers and you'll have cracked it.

Mastering a digital clock – in practice

Before you get too excited and volunteer to change every digital clock in the house, it is only fair to point out that the chances of this panning out are roughly on a par with those of a pantomime horse being elected the next leader of the Conservative Party. A more likely sequence of events is this:

❶ All or part of the display will start flashing – and once the numbers start flashing, there is no stopping them.

❷ You will need a lie-down to recover and regain your composure.

❸ If you are lucky, the flashing number is a prelude to a change of setting. By pressing the button repeatedly, you may just be able to advance the display to the required time.

❹ You will accidentally press the button for too long and have to go all the way round again.

❺ You will threaten to throw the whole thing out of the window and invest in a sundial and cockerel instead.

Another sensible option, of course, is simply to leave all your clocks on GMT and leave Post-it notes lying around reminding you to add one hour whenever you check the time.

Video and DVD recorders

The advent of digital TV with its myriad channels has made a recorder more essential than ever. After all, if you're glued to an umpteenth repeat of *Are You Being Served?* you'll be missing out on an entire day's programming devoted to *Diagnosis Murder*, as well as live darts from a working men's club in Barnsley. It simply won't do.

If you don't set the clock correctly on your recorder, you have only yourself to blame when the inevitable occurs: you either miss the first five minutes of the show you are trying to record, or, worse, you miss the last five. Few television experiences are more frustrating than sitting through 115 recorded minutes of *Midsomer Murders*, only for the screen to dissolve into a commercial for verruca gel just as Barnaby is about to reveal the killer's identity.

Of course, if the clock *is* correct on your video or DVD recorder, Sod's Law dictates that you will get the timing right but record the wrong programme entirely. Instead of the FA Cup replay, you'll be stuck with a documentary about the history of Velcro.

Power cut

This is the ultimate betrayal. Digital clocks promise so much and look so competent, but a momentary loss of power makes them go haywire. Not only do they instantaneously forget all your radio settings, but they decide, in their madness, that it must be midnight and reset to 00:00 (usually flashing, for extra annoyance).

This is particularly inconvenient with alarm clocks, as a power cut during the night will result in the clock failing to trigger the alarm in the morning. So you oversleep and end up having to answer the door in your dressing gown even though, as the old joke says, not many people have a door in their dressing gown.

Advanced digital-clocking

If you're reading this thinking 'Pff – it's a doddle', just have a crack at taming the digital clock on your central heating and hot water programmes. Go on – I dare you. Failure to master the 'override' and 'advance' buttons can result in you ending up with enough hot water to thaw Antarctica, or a central heating system that only operates between 2am and 5am every Tuesday.

BANKS WITH REAL PEOPLE IN THEM

One of the biggest complaints about banking today is that nobody is interested in the customer. With online banking, this is not the case. The moment you log on, the chances are that someone somewhere is very interested in you and is eagerly trying to hack into your account.

New-fangled online banking: Pros and cons

- ✓ No need to queue or deal with a growing line of irritated people behind you.
- ✗ Unless you have teenagers in the house.

- ✓ Easy to use.
- ✗ You have to remember your online banking membership number, PIN and password. At a proper bank you barely need to remember your own name.

- ✓ Easily accessible at all hours.
- ✗ If the extent of your home technology is a portable typewriter, online banking is no more accessible than the summit of Everest by mobility scooter.

Old-style banking: Not all it's cracked up to be

Before we get too carried away in our ode to proper banks, let us not forget that they, too, are fraught with annoyance.

❶ Banks seem to have fewer branches than a Norfolk family tree. So many have been closed down and converted to alternative premises – because there just aren't enough places for young people to get drunk or buy mobile phones.

❷ Banking law appears to stipulate that only one of the eight cashier windows should be open at any given time. This is not to say that there's a shortage of staff: there are dozens of them! They're just too busy filling out paperwork – probably their lottery numbers – while desperately trying to avoid making eye contact with the increasingly rage-filled customers.

❸ They try to force you to use machines in any case. You know, those complicated cheque-deposit letterboxes that require you to throw your hard-earned cash into a hole in the wall. But who knows what sort of unscrupulous individual might be lurking on the other side of the slot? No, no, no: it's much safer to give it to someone with a face that you might be able to pick out in an identity parade should the need arise. Unfortunately the majority of customers feel the same, with the result that the length of the cashier queue makes you wish you had brought your sleeping bag.

Which brings us neatly onto...

Queuing

Even though the British have turned it into a national pastime, queuing can be immensely frustrating. There are many horrendous things about being trapped in a queue, but these are the worst of them:

The happy couple: The inevitable glee-filled – if not positively giddy – people on the posters lining the walls, beaming from ear to ear because the bank has just granted them a £10,000 loan. Why are they so happy? Don't they know they'll need another loan just to pay off the interest? Why don't banks put more realistic posters on the walls, such as a couple looking on helplessly while their home is repossessed?

The dreadful Howard. You know, the guy from the Halifax commercials. Some branches even have a full-sized cardboard cut-out of him, as if he were a mystical eastern deity rather than a daft bloke from Birmingham in thick glasses.

The view. When you have exhausted all the promotional offers and managed somehow to ignore the fact that you are being coughed over by the sickly person behind you, your attention inevitably turns towards the bank's security mirror in which you can't help noticing that the bald patch on the top of your head has grown bigger – and that's just since you started queuing.

Queue-rage control

Good things to think of while queuing:
A woodland carpet of bluebells
A gentle waterfall
Lambs gambolling in a field

Bad things to think of while queuing:
Whether you remembered to switch the
    oven off before coming out
Rising interest rates
The time

Success at last

Even when you make it to a window and realize with considerable relief that it wasn't just a mirage all along, your ordeal is not over. You will be asked all manner of security questions, from the balance of your account to your inside leg measurement. You can't help feeling that if they thought to ask the same list of questions when anyone attempted to rob the place, they would be able to detain the culprit until the police arrived.

One school of thought considers that having spent anything up to half an hour in line, you have earned the right to take your time when you finally reach the window and experience human contact. So feel free to chat to the cashier about your holidays, your back trouble, and Mrs Milkins at number 42. Don't worry if you struggle with this at first: you will find it comes naturally the older you get.

ELECTRONIC GARDEN GADGETS

If the nearest thing in your shed to a power tool is a trowel and if your lawn mower first saw action during the Boer War, your garden will be the relaxing, welcoming oasis of calm that it should be.

If you've fallen prey to modern electrical gadgets, however, the pleasant song of the blackbird will have been drowned out by the buzzing and grinding of strimmers and trimmers, their screeching sound broken only by the occasional anguished cry of pain followed ten minutes later by the whine of ambulance sirens as you or your nearest and dearest are rushed to casualty.

Garden wisdom

Since most of us prefer to come in from the garden with the same number of fingers we went out with, supposed labour-saving devices must be treated with caution and respect, especially in the hands of the inexperienced or the downright incompetent. Follow these tips to gain maximum satisfaction and minimal loss of limb:

> When operating a chainsaw, it is essential to wear the correct safety equipment. A full suit of armour should suffice.

> Even if your razor has broken and you have an important black-tie function to attend, never be tempted to give yourself a shave with a hedge-trimmer.

> Don't get too carried away with a chainsaw. Even to the amateur horticulturist there is a subtle difference between pruning a shrub and reducing it to within an inch of its life. As a general rule, if no growth is visible above ground level, you have gone too far.

❯ Resist the temptation to recreate a scene from *The Texas Chainsaw Massacre* on your neighbour's *leylandii*.

❯ Before activating a hedge-trimmer, check that there are no minor obstructions in your path... such as telegraph poles, street lamps or the postman.

Gardening facts

❶ Lawnmowers are just plain dangerous.
People have been known to lose shoes, toes, even
entire lawns, through encounters with rogue
mowers. If the blades don't get you, the electric
cable will, as you go to such lengths to keep the
cable away from the blades that you move in
ever decreasing circles until you manage to truss
yourself up like a joint of lamb and need to be
rescued.

**❷ You will look like a twit astride a ride-on
mower if your garden is little bigger than a
picnic blanket.** Instead, use a simple push mower,
and treat it as a walking frame that cuts grass.
Indeed, for the price of a pair of shears, some rope
and a set of castors you could probably customize a
walking frame to do the job.

❸ **Your garden will not automatically grow in immaculate stripes, contrary to the photo on the grass-seed packet.** Simply don't bother trying to achieve a stripy lawn. All it takes is one detour round a molehill and the rest of the stripe will disappear beneath an assortment of dandelions, plantain, daisies, etc. With dark patches of clover and hop trefoil also vying for space, the overall effect becomes more polka dot than parallel lines.

Should you have a lawn the size of a field with areas of long, untamed grass, either buy a petrol mower or a goat. The latter can be just as temperamental and dangerous as a motor mower but its cheese is immeasurably better than that from a Qualcast.

RELIABLE TRADESPEOPLE

As you get older, some things become harder than ever to find: your keys, your slippers, your house... Reliable trades-people fall into the same category, and so finding one who will keep an appointment and do a decent job without hideously overcharging you is like stumbling upon gold dust.

Three facts

❶ Workmen will never turn up on time.
Never mind the same day, you're lucky if it's the same month. It makes you wonder what happens when builders get married. Do they turn up at the church at ten o'clock on Monday morning and say, 'Sorry, love, I was getting married to someone else on Saturday – last-minute rush job, too good to refuse. I thought it would only take an hour, but I just couldn't get away. You know how it is.'?

❷ They never have the right tools for the job.
This means they will have to go away for another hour or two to collect the right tools. In these circumstances, it is always advisable to hold one of them hostage in case you never see the others again.

❸ The job will always be 'bigger than expected'.
This is a roundabout way of saying 'more expensive', and will be accompanied by a shake of the head, and some faux-reluctant mutterings about 'the idiot who did this'. Unfortunately, you are that idiot.

Lots more rules

It's better to be safe than sorry, eh?

> Never commit yourself on the spot to a cold caller who rings your doorbell unannounced, claiming that his colleagues are working on another job in the area... particularly if he is actually wearing a cowboy hat.

> Resist the temptation to pick the first firm listed alphabetically. 'Aardvark Plumbing' may well look good at the top of the list but remember: an aardvark is a termite-eating nocturnal mammal with a tubular snout and a long extensible tongue. It is not an expert in central heating systems or guttering repairs.

❯ Never trust any workman who hides behind the sofa when a police car drives past the house.

❯ Never trust any workman who actually looks as if he has fallen off the back of a lorry.

❯ Never hire a manual worker who paints his finger nails and wears eyeliner. His heart won't be in it.

❯ Never pay cash upfront. At best, they'll do a shoddy job; at worst, they'll do a runner.

Tradesperson dictionary

When leafing through *Yellow Pages*, it is easy to be seduced by the promises made in the advertisements. However, many are less appealing when translated:

'No job too small'	'We're desperate'
'Local specialists'	'We live nearby'
'Local authority approved'	'We did the mayor's house once'
'All areas covered'	'No builder's bottom'
'Over forty years' experience'	'We're a bunch of old men, so no heights and we need a nap in the afternoon'
'All work guaranteed'	'You can take us to court'
'OAP reductions'	'We're desperate'
'A name you can trust'	'...to muck things up'
'Prompt and courteous attention'	'We've got an answering machine'
'Fully insured'	'We have a lot of accidents'
'Skilled and time-served craftsmen'	'Our skilled craftsmen have served time'
'No obligation'	'...but don't walk home alone in the dark'
'Honest reputation for over twenty years'	'We haven't been caught yet'
'We will beat any other quotation'	'We're desperate'

BUYING AND SELLING ON EBAY

Ten years ago, if there was a major disaster, our first thought was: 'how can we help the survivors?' Now whenever there's a major disaster, our first thought seems to be: 'can we sell bits of the wreckage on eBay?'

> The very first item sold on eBay was a broken laser pointer for about £8. When the seller contacted the winning bidder and asked if he realized that the laser pointer was broken, the buyer replied: 'I'm a collector of broken laser pointers.'

You can get just about anything on eBay, although there are some prohibited items:

> Firearms and ammunition
> Human body parts
> Used underwear
> Lottery tickets
> People

Everything else is fair game, including toenail clippings, half-eaten food, spat-out chewing gum, dirty socks, earwax – they're all considered sought-after items if they once belonged to a celebrity. Some Hollywood waiters are so keen to cash in that they take away famous customers' meals before they have finished just so that the remnants can be sold on eBay. In the strange world of online auctions a half-slice of toast bearing the imprint of Justin Timberlake's teeth creates almost as much excitement as a Rembrandt.

eBay: an inventory

Here are some of the more unusual items that have been offered for sale on eBay:

> Wind 'captured' from Hurricane Frances and stored in Tupperware containers: £5 a pop

> An Indiana woman's dead father's ghost, in the form of the deceased's walking-cane: £34,500

> A half-eaten toasted cheese sandwich apparently bearing the image of the Virgin Mary: £15,000

> A bottle of Lake District air: £60

> A ball from Elvis Presley's pool table: £900

> British ex-pat Ian Usher's entire life in Australia, including his home, car, job and friends, all of which reminded him of his ex-wife: £200,000.

❯ Water supposedly left in a cup that Elvis Presley drank from in 1977: £230

❯ The entire town of Bridgefield, California: £1 million

❯ A single cornflake previously owned by a Coventry University student: £1.20

❯ A Salt Lake City woman's forehead as the site for a permanent, tattooed advertisement: £5000

❯ A bucket of Bournemouth seawater: £50

❯ An Arizona man's air guitar: £3

❯ British disc jockey Tim Shaw's £25,000 Lotus Esprit sports car: a 'Buy It Now' price of 50p. It was put up for sale by his wife after she heard him flirting with model Jodie Marsh on air. Hell hath no fury like a woman scorned with access to eBay.

FAQs

Is this the genuine article? An excellent question. Who's to say that the lock of hair you've just bought is in fact Madonna's? A man from Brisbane once tried to sell New Zealand at a starting price of Aus$0.01. The price had risen to $3000 before eBay wisely closed the auction.

What if the product description is misleading? Always treat items for sale with a degree of scepticism, steering clear of things such as 'Elvis Presley's digital camera' or 'Early Roman telephone directory'. With others such as 'King Charles's spaniel', it is probably worth double-checking the punctuation and finer details before entering a bid.

How do I know I'll ever receive the goods? Money disappears from your account and you spend a fraught few days panicking about whether your item will actually arrive – and in what state. Usually your only contact for the seller is an email address, but what you really want is a home phone number, home address, and their car as security.

It feels rather like winning a prize when you end up as the top bidder on an item: 'you have won this item!' you are told – when really it's less a case of 'winning' than of 'being contractually obliged to pay for' something.

VIDEO GAMES

The main problem with video games is that you have to do everything so quickly and perform so many operations simultaneously. Our brains might have been able to cope with such demands when we were younger, had video games been around then, but now anything more taxing than a Junior Sudoku puzzle requires a lie-down.

To watch youngsters at the controls of their PlayStations and Nintendos, however, is to witness a digital frenzy, as fingers and thumbs co-ordinate at lightning speed to negotiate a race track, a perilous cave or a horde of killer aliens – or, more likely, a horde of killer aliens racing each other through perilous caves.

It's just plain embarrassing

Your most cringe-worthy performances will doubtless be on games at which you should logically be able to beat a small child, such as driving. But try any driving game and you will find that you:

a) Drive in the wrong direction until the computer overloads and crashes.

b) Take 140mph corners at a sensible 10mph, as the other cars flash past for the fourth time.

c) Run an ongoing commentary along the lines of 'This is a bit fast, isn't... hey! Where did that come from? Do I press the red button to... Oh no, what have I done now?' And so on.

d) Generally look as if you are tootling off to the coast on a Sunday afternoon rather than taking part in a Grand Prix.

Five facts about video games

❶ You will be no good at any of them. If you are, you have no right to call yourself a Senior.

❷ Golf will turn out to be the worst traitor. Like driving, golf is something you are supposed to be good at, having perhaps done it in real life. But no: your tee shot will come to rest either two feet in front of you or twelve miles to your right, and you'll spend more time in bunkers than Hitler.

❸ You may well have spent your formative years playing air guitar along to Led Zeppelin records and perhaps you once saw Robert Plant at the farmers' market, but don't fool yourself that you are a Guitar Hero.

❹ It is best to steer clear of bowling or darts games
– you are liable to forget where you are and send
the controller hurtling at high speed towards the
TV screen.

❺ The on-screen commentator will go from
'Oh dear, that wasn't very good, was it?' to
complete silence, his vocabulary to describe
ineptitude thoroughly exhausted.

While all of this is going on, your junior competitor will
complete the game in 'expert' level and run off to find
someone fun to play with.

DRESSING YOUR AGE

A probable clothing timeline of your life

> **Mid twentieth century: your childhood**
> For boys, depending on which decade you were
> raised, it was a chaotic mess of winkle pickers, tank
> tops, flared trousers and kipper ties. Meanwhile
> girls were forced to parade around in platform
> heels, hot pants and puffball skirts, more often than
> not topped with either a beehive hairdo or a tight
> perm. It is the stuff of which nightmares are made.

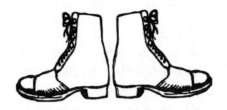

> **Later twentieth century: your adulthood**
> At last you had reached an age where you were
> no longer expected to be dedicated followers of
> fashion, and you breathed a huge sigh of relief.
> You started wearing comfortable clothes with the
> vaguest of nods towards the trends of the day.

> **Late 1990s: Tony Blair is pictured wearing
> jeans. Worse still, William Hague dons a cap.**
> You realized with horror that middle-aged people
> don't always look good in jeans, particularly the
> figure-hugging sort that strive so manfully to cling
> to the contours of Jeremy Clarkson.

> **Now: your seniority**
> If you are anything over fifty, presumably this
> means wearing a floral frock and court shoes, or a
> tie with soup down it and beige trousers with urine
> stains.

Dressing your age

If you try to look too trendy in your forties, fifties and beyond, you are told bluntly by the younger element, 'Don't you think you should be dressing your age?' Or that you look like mutton dressed as lamb. It could be worse: you could be mutton dressed as mutton.

But what should you wear? Of course not all young fashions are suitable for later life – such as ripped jeans. Come to think of it, they're not suitable for wearing at any age. When we were young, if our jeans were torn we patched them up or threw them out; nowadays if the jeans aren't torn, the kids rip them deliberately. It's like buying a new car and taking a pickaxe to smash a gaping hole in the wing.

Otherwise there is no reason why mature people should not be able to adapt some of today's fashion trends to their tastes, as long as they obey certain rules:

For him:

❯ No matter how much excess baggage you are carrying in the stomach area, it is never a good look to wear trousers whose waistband comes up to your armpits.

❯ Trainers are a wonderful comfort for ageing feet, but resist the temptation to wear them with a suit at your daughter's wedding.

❯ It is acceptable to wear your shirt hanging out of your trousers. It can hide a multitude of sins.

❯ There is no excuse for sandals worn with socks, nor for very short shorts.

❯ Hooded tops should be worn only when jogging, i.e. not at all.

For her:

> Puffa jackets and very short skirts should be avoided unless you're actively looking for business.

> Leather trousers are not appealing on a bottom of such size that it still looks as if they are being worn by the cow.

> If you are worried about your ageing neckline, try wearing polo neck tops – or a neck brace.

❯ There is no excuse for leopard print.

❯ Don't wear cropped tops that show your navel if you're more than a size 12, unless you want to act as some kind of deterrent.

❯ Pay no attention to self-styled fashion gurus Trinny and Susannah, the only double act where one half looks as if she is about to eat the other.

BUDGET AIRLINES

If you are used to reclining in first-class with a complimentary brandy and copy of *The Times*, budget travel is a different world. On a budget airline, the only free drink you will get is if somebody spills their tea over you, while the available reading material amounts to the in-flight safety procedure card.

There may be trouble ahead...

Most companies are thoroughly reputable but there are
certain telltale signs that you may have booked with a bad
budget airline:

> The seat-belt sign doesn't go on before take-off in
case the passengers have to get out and give the
plane a push start.

> Instead of the safety drill, the flight attendants get
the passengers to practise synchronized screaming.

> Just before take-off, one of the crew announces to
the passengers, 'OK. It's time to elect a pilot.'

> When the steps are pulled away, the plane leans to
that side.

> You hear the pilot say, 'Let me get this straight. If I
push the lever this way, the plane goes up?'

> Turbulence is the in-flight entertainment.

Where am I?

The first thing to consider when flying by a budget airline is not necessarily the price but the destination. You need to check that your flight to Milan does actually land at an airport somewhere near Milan and not one that is closer to Rome. Vienna, for instance, has two airports: one that is in the environs of Vienna, as one might expect – and one that is in Bratislava, the capital city of neighbouring Slovakia.

So beware: if the suggestions for getting from the arrival airport to your destination involve three days of travel (partly by camel) or a taxi fare equal to the national debt, choose another airline.

What not to pack

Never mind foregoing all liquids, razors and shoes in these security-conscious times; and never mind debating whether you really need six pairs of trousers for a weekend in Paris or a sheepskin coat for the Costa del Sol in June ('because it can turn chilly in the evenings'). No: the simple rule is not to take any luggage whatsoever, since a budget airline will charge you an exorbitant amount for the privilege. You may take one small book and a pair of socks and call it 'hand luggage'.

If you can live for two weeks on the contents of your pockets, budget airlines are for you.

The 'budget scramble'

Budget airlines don't allocate seat numbers, but they do have a marvellous scam called 'priority boarding'. For a fee (of course), you can be first in the queue to board the plane, along with all the mothers and children. The problem is that most budget airlines have to park their planes miles away from the terminal building, so your super-duper priority pass will bag you a great seat on the shuttle bus – thereby ensuring that you are stuck at the back of said shuttle bus while the cheapskates wedged in the doorway dash off to grab the best seats on the plane.

Save yourself the money and simply get to the departure gate in time to get onto the first busload of passengers. There is plenty to do during your cramped wait at the gate:

❶ Watch with awe as your brand new cases are hurled and bounced onto the plane by the trained handlers. It is at such times that you are eternally grateful that they chose baggage handling as a career rather than gynaecology.

❷ Put your brain to the test trying to work out how on earth the eight-tiered queuing system works.

❸ Marvel at the speed with which normally mild-mannered souls – pillars of the community, devoted charity workers without a blemish on their driving licences – are suddenly transformed into crazed monsters in their bid to be one of the first to board the plane. Any lone traveller with a stoop and a walking stick is an obvious target for queue jumpers and may be dragged unceremoniously through the gate by complete strangers before being dumped on the tarmac while the kidnappers dash off to complete the race for the best seats on the plane.

Decision time

There is only ever a split second in which to make the biggest decision of all: front or rear steps? At the front you're more likely to exit the plane first and take up the best position at luggage claim; at the back you're more likely to survive should the worst happen. It's the modern-day traveller's greatest dilemma.

You can guarantee that whichever door you choose will be the wrong one, because you will invariably be stuck behind someone who persistently fiddles with the overhead compartment. First he puts his bag up there, then he realizes he wants something from it. But no sooner has he started to sit than he decides to take off his coat and put that up there, too. Then he remembers something else, and so on. In the space of a minute he is up and down more than the FT Index. By the time he has exhausted this routine, the only seat left is barely visible between a heavily pregnant woman and an obese chatterbox.

Rest and relaxation

When it comes to rest and relaxation during your flight, budget airlines are no different to their more expensive cousins, i.e. there is none.

As soon as the snack trolley has passed, an attendant comes round with the company magazine, followed by the lucky scratch cards, the duty-free trolley, the snack trolley revisited, the waste paper collection, the company magazine collection, and the offer of cut-price hotel vouchers. On a short flight your attention is demanded approximately every five minutes, meaning that there is about as much chance of falling asleep on a plane as on the central reservation of the M25.

In conclusion

In truth, as long as your expectations aren't high and your plane is, flying by budget airline is no worse than any other form of air travel. Indeed it's considerably quicker than a hot air balloon and very nearly as comfortable as a hang glider.

But pack a pilot's uniform in your hand luggage – just in case.